When the Road Price Is Right

Land Use, Tolls, and Congestion Pricing

Urban Land Institute

Infrastructure Initiative

About ULI

The mission of the Urban Land Institute is to provide leadership in the responsible use of land and in creating and sustaining thriving communities worldwide. ULI is committed to

- Bringing together leaders from across the fields of real estate and land use policy to exchange best practices and serve community needs;

- Fostering collaboration within and beyond ULI's membership through mentoring, dialogue, and problem solving;

- Exploring issues of urbanization, conservation, regeneration, land use, capital formation, and sustainable development;

- Advancing land use policies and design practices that respect the uniqueness of both the built and natural environments;

- Sharing knowledge through education, applied research, publishing, and electronic media; and

- Sustaining a diverse global network of local practice and advisory efforts that address current and future challenges.

Established in 1936, the Institute today has nearly 30,000 members worldwide, representing the entire spectrum of the land use and development disciplines. ULI relies heavily on the experience of its members. It is through member involvement and information resources that ULI has been able to set standards of excellence in development practice. The Institute has long been recognized as one of the world's most respected and widely quoted sources of objective information on urban planning, growth, and development.

Recommended bibliographical listing: Urban Land Institute. *When the Road Price Is Right: Land Use, Tolls, and Congestion Pricing*. Washington, D.C.: Urban Land Institute, 2013.

ISBN 978-0-87420-262-5

About the ULI Infrastructure Initiative

The mission of the ULI Infrastructure Initiative is to promote more sustainable infrastructure investment choices and to foster an improved understanding of the links between infrastructure and land use. Because infrastructure is the foundation for metropolitan prosperity, and because it provides the physical framework for real estate investment, ULI has identified infrastructure as a key priority.

Established in 2007, the initiative achieves its mission through a multifaceted program of work that leverages ULI's extensive public and private networks and includes research and education, publications, and convenings.

The ULI Infrastructure Initiative is led by full-time staff with a deep understanding of global infrastructure challenges and opportunities. An advisory group composed of industry leaders guides the program. Key Infrastructure Initiative activities include

- A global infrastructure report, produced in collaboration with Ernst & Young since 2007;

- Ongoing programming to promote regional infrastructure solutions;

- Regular e-mail updates to infrastructure audiences; and

- Exploration of infrastructure topics relevant to ULI members and to ULI's public and private partners.

Connect with the ULI Infrastructure Initiative

To learn more about the ULI Infrastructure Initiative's work, visit the website (www.uli.org/infrastructure) and blog (www.uli.org/infrastructureblog) and follow us on Twitter (@uli_infra). To subscribe to the *Infrastructure Update* newsletter, e-mail infrastructure@uli.org.

The ULI Infrastructure Initiative welcomes new partners and sponsors. To explore opportunities to work with us, contact infrastructure@uli.org.

Project Staff

Rachel MacCleery
Vice President, ULI Infrastructure Initiative

Sarah Jo Peterson
Research Director, ULI Infrastructure
    Initiative

Casey Peterson
Project Employee, ULI Infrastructure
    Initiative

Production Staff

James Mulligan
Managing Editor

David James Rose
Manuscript Editor

Betsy VanBuskirk
Creative Director

Anne Morgan
Graphic Designer

Craig Chapman
Senior Director, Publishing Operations

Authors

Sarah Jo Peterson
Research Director, ULI Infrastructure
    Initiative

Rachel MacCleery
Vice President, ULI Infrastructure Initiative

Contents

Figures

Abbreviations

BRT   bus rapid transit

HOT   high-occupancy/toll

HOV   high-occupancy vehicle

ICC   Intercounty Connector

IHS   Interstate Highway System

TOD   transit-oriented development

Executive Summary

The United States is in the beginning stages of a significant shift in how the country collects revenue from—and pays for—its transportation system. After decades of depending on taxes on motor fuels, federal, state, and metropolitan leaders are expanding the use of tolls. They are looking to tolling to pay for new and revitalized transportation infrastructure. They are also experimenting with tolling's powerful ability to provide a new benefit to road users: the reliable, congestion-free trip.

Compared with other potential revenue sources, such as sales taxes, income taxes, and even taxes on motor fuels, tolling and related schemes to charge a tax or fee for every mile driven have much stronger potential to affect decision making about land use. Tolling that enables congestion-free travel increases the probability that the tolled road will have impacts on land use and development.

New types of tolling also present decision makers with many choices about how to manage new tolled roads and how to coordinate tolled facilities with transit service. The outcome of these decisions will influence how these new facilities interact with land development and whether the opportunities they promise for widespread benefits will be fully realized.

ULI believes it is important to include land use—the impacts, concerns, and opportunities—in discussions about the future of tolling and other changes to how revenue is collected from transportation. The prism of land use should encompass how tolling's impacts on land use will affect societal equity.

To raise awareness of land use and frame the issues for future research and discussion, the ULI Infrastructure Initiative convened a group of experts, including leaders representing transportation and land development, to participate in workshops and interviews conducted in the summer of 2012. Study participants explored scenarios constructed as part of a thought experiment on how land development in metropolitan areas would be affected by the spread of congestion pricing, the expansion of tolled highways, or the adoption of taxes on vehicle-miles traveled (VMT).

The ULI Infrastructure Initiative also developed five brief case studies of tolling in Florida, Texas, California, Colorado, and Maryland. The case studies illustrate the different policy options for managing travel reliability, traffic volume, travel speeds, and revenue targets and for integrating tolling and transit service. They also explore how these new tolled facilities are being coordinated with land use and development.

This report documents the results of the ULI Infrastructure Initiative thought experiment and the case studies. Among the most significant conclusions:

- The potential for tolling and other new transportation revenue mechanisms to influence land use is real, but the magnitude of the impacts is likely to be modest.

- The most dramatic impacts—and opportunities—are likely to be located in the corridors surrounding tolled roads or highways with the option for congestion-free travel.

- The impacts on land use will vary greatly by metropolitan region and will be influenced by the transportation network—including mass transit services—land use patterns, local land use policies, and economic trends.

- Tolling and other transportation revenue mechanisms have the potential to interact with land use in ways that support growing market preferences for development in compact, mixed-use, walkable nodes, but achieving this objective will require careful coordination with land use policies and other transportation services, include transit service.

- Because tolling that manages congestion is permanent and not priced according to the cost of the facility, policy discussions need to include the appropriate uses for "excess" revenue.

Developers, planners, and researchers should work within their respective professional communities to advance and disseminate knowledge about these new uses for tolling and congestion pricing and their impacts on land use. Reaching out across disciplines will strengthen efforts to conceptualize research projects, develop best practices, and set standards. Of equal importance, developers, planners, and researchers offer unique perspectives that will be valuable to policy makers at the federal, state, and local levels. All should strive to ensure that these new transportation services and the land uses that are attracted to them achieve a broad set of policy goals without creating unintended consequences.

By paying attention now—through framing the issues for policy discussions, future research, and the development of best practices—Americans can more fully realize the potential opportunities to tie transportation choices to desired land use outcomes.

PART 1:
Shifting Approaches to Collecting Revenue from Transportation

Imagine living and working in American metropolitan areas 20 years from now:

> **Will the Interstate Highway System be converted to toll roads, making expressway travel available only for an extra charge?** To what lengths will people go to avoid the toll roads? Will locations near the toll roads be attractive places to live? To work? To shop?

> **Will the government send you a monthly bill charging for every mile you drive?** Or will you opt to load miles in advance on a card that triggers a meter installed in every car? Will you think about that charge per mile every time you decide on a destination and make a vehicle trip? Or, instead, will you prefer to pay the charges through an unlimited-miles plan?

> **Will the stress of being stuck in congestion on highways be a thing of the past?** Instead, will people talk about choosing to use the free-flowing express lanes—by riding a bus or paying a higher toll for a trip by personal vehicle—or taking their chances in the regular lanes? How much of a premium will people be willing to pay for locations near the best congestion-free express lanes?

> **Will the new toll roads, charges per mile of driving, and the option for congestion-free highway travel change how people think about land use?** Will they change where people want to live? Locate a business? Shop and recreate?

The United States is on the cusp of a tremendous shift in how the country collects revenue from—and pays for—its transportation system. After decades of depending on the proceeds from taxes on motor fuels, decision makers are looking anew at a wide range of revenue sources to fund investments in and operations of roads, bridges, and mass transit.

In search of revenue to pay for transportation infrastructure, states and metropolitan areas are increasing their use of tolling, both on new roads and on additional lanes along existing roads. They are starting to charge tolls for access to carpool lanes, as a way to better use capacity and raise revenue, and as a way to provide congestion-free travel options. States and metropolitan areas are also experimenting with technologies that make it possible to charge for each mile of driving.

Compared with other revenue sources, such as sales taxes or income taxes, what makes tolls and charging per mile different is their much stronger potential to affect decision making about land use. Tolls that enable congestion-free travel increase the probability of impacts on land use.

ULI believes it is important to include land use—the impacts, concerns, and opportunities—in research, discussions, and debates about the future of tolls and other transportation revenue mechanisms, including taxes on motor fuels and charging per mile. *When the Road Price Is Right* is part of a ULI Infrastructure Initiative project designed to raise awareness of land use in debates at the federal, state, and local levels about how Americans pay for transportation. By paying attention now—through framing the issues for policy discussions, future research, and the development of best practices—Americans can more fully realize opportunities to tie transportation choices to desired land use outcomes.

When the Road Price Is Right contains three parts. Part 1 describes changes in how Americans pay for transportation, the types of revenue sources under consideration, and why tolls and proposals that charge per mile have stronger potential to trigger a land use response than other revenue mechanisms. Part 2 presents the results of a series of ULI-led workshops and interviews with leading experts in transportation and land development who analyzed the land use impacts of charging tolls, collecting a tax on every mile driven, and instituting value pricing to provide free-flowing traffic on expressways. Part 3 covers conclusions for today's decision makers and recommendations for planners, developers, and researchers.

Taxes on Motor Fuels: A Declining Revenue Source

For decades, federal and state taxes on motor fuels have funded the lion's share of surface transportation infrastructure investments. Moreover, there was a rough relationship between the amount of revenue generated and transportation demand. Revenue increased the more people drove their cars and trucks and the more fuel their vehicles consumed. Because revenues from federal and state taxes on motor fuels are dedicated (for the most part) to spending on surface transportation, more and more driving —the trend throughout the second half of the 20th century—meant more and more revenue for roads, bridges, and mass transit.

American drivers, however, did not have to give the tax much thought. As an excise tax charged per gallon of fuel and collected from fuel suppliers, the tax was—and is— indirect and mostly hidden.

Starting in 2000 or so, the reliability of revenue from motor fuel taxes began to decline. Americans stopped increasing the amount they drove; vehicle-miles traveled per capita peaked in 2004 and even total vehicle-miles traveled flattened out by the end of the decade. The mid-decade run-up of fuel prices sent Americans in search of fuel-efficient vehicles and other transportation alternatives, decreasing even further both fuel consumption and vehicle-miles traveled for passengers and freight.

....

Figure 1: VEHICLE FUEL EFFICIENCY AND MOTOR FUEL TAXES

When Fuel Efficiency Increases, the Revenue Collected from Motor Fuel Taxes Declines

Federal excise tax on gasoline	18.4 ¢ per gallon
State excise taxes on gasoline*	21.0 ¢ per gallon
TOTAL	**39.4 ¢ per gallon**

Vehicle fuel efficiency (miles per gallon)	**Tax rate per mile**
10 mpg	3.9 ¢ per mile
20 mpg	2.0 ¢ per mile
35 mpg	1.1 ¢ per mile
50 mpg	0.8 ¢ per mile

* Average of state excise tax rates on gasoline as calculated by the American Petroleum Institute.

....

Forecasters warn that these changes in American fuel consumption are not just blips in the long-term trend or echoes of the recession. Rather, they foretell a new future—one in which Americans drive more fuel-efficient vehicles fewer miles, and where taxes on motor fuels never recover their former power to generate stable—let alone increasing—revenues.

Despite the declining burden that the tax rate on motor fuels imposes on taxpayers, little political support has materialized for increasing it to make up for falling consumption or to meet current and future infrastructure needs. The U.S. Congress last increased taxes on motor fuels in 1993, and only 15 states raised their tax rates on motor fuels between 1997 and 2011. The diminishing support for continuing to pay for surface transportation with revenues from taxes on motor fuels became apparent at the federal level in 2005, when Congress passed SAFETEA-LU (the Safe, Accountable, Flexible, Efficient Transportation Equity Act: A Legacy for Users), a major long-term surface transportation bill, knowing that federal taxes on motor fuels would not generate enough revenue to cover the bill's authorized spending. In 2008, Congress had to step in with general revenues for the first "bailout" of the Highway Trust Fund. In 2012, Congress gave up on passing another long-term surface transportation bill. SAFETEA-LU's replacement, MAP-21 (the Moving Ahead for Progress in the 21st Century Act), is a two-year bill that caps spending at existing levels. Even with these limitations, Congress still had to top off the funding with general revenues.

Although per-gallon tax rates that do not account for inflation, reductions in driving, and increases in fuel efficiency mean that the tax burden posed by taxes on motor fuels has declined since 1993, U.S. households are still paying more for transportation overall. A recent study by the Center for Housing Policy and the Center for Neighborhood Technology estimates that between 2000 and 2010 transportation expenses increased by 33 percent, but household income increased by only 25 percent. The continued trend of increasing transportation costs complicates government efforts to raise additional funds via taxes on motor fuels or other transportation-related revenue mechanisms.

Technology Breathes New Life into Tolling

At the same time that transportation policy makers scramble to supplement revenues from taxes on motor fuels, advances in communication and information technologies allow for entirely new ways of thinking about the oldest of transportation revenue mechanisms—tolling.

The revolution in tolling is spreading. First came the electronic toll systems that made it possible to travel on toll roads without stopping at toll booths. Then came the high-occupancy/toll (HOT) lanes that allowed drivers—for a price—to take advantage of the free-flowing traffic in the carpool lanes that line many of the freeways in congested metropolitan areas. These two examples are just the tip of the iceberg when it comes to the different ways new communication and information technologies can be used to better manage transportation demand and raise additional revenue.

What makes these new technologies so powerful is the ability to vary the charge for driving based on the type of roadway and the time of day. Variable pricing allows for value pricing—charging the customer based on how much he or she values the service, instead of based on the cost to provide the service. Value pricing can be used to give the customer options such as paying more for faster travel. Value pricing can also be used to manage capacity, reducing demand for expensive projects that add new lanes.

Under certain conditions, value pricing has the potential to generate revenue significantly higher than what is needed to build, operate, and manage the transportation facility. It is possible that revenue from value pricing could fund general transportation needs and a range of other benefits.

Connections between Land Use and How to Pay for Roads

Transportation and land use are inextricably linked. The old joke about real estate—"the first three rules of real estate are location, location, location"—is a reflection of the reality that the value of a parcel of land is largely determined by its ability to connect with other desirable parcels of land. Exceptions to this rule are few and far between.

Land value is affected by decisions the public sector makes about transportation investments. If the government decides to build a new high-speed road from the urban edge to outlying areas, for example, the land served by the roadway gains value because it is closer—in terms of travel time—to desirable locations.

Figure 2: EXPRESS LANES AND HOV LANES IN THE UNITED STATES

Miles of HOT/Express lanes in the United States are rapidly expanding; under federal law, existing HOV lanes are now eligible to be converted to HOT lanes.

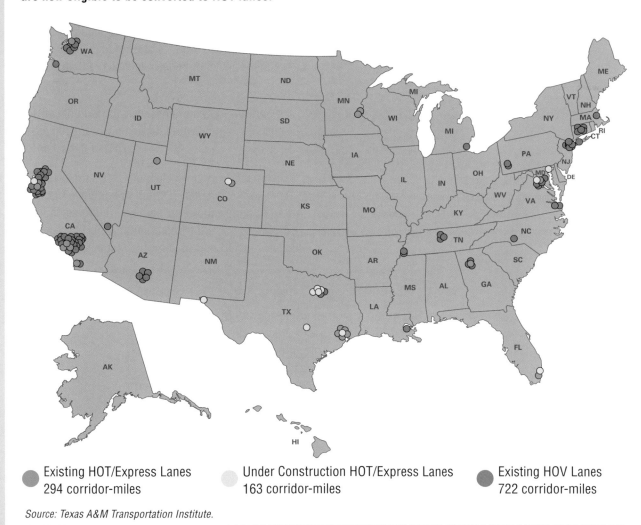

● Existing HOT/Express Lanes
294 corridor-miles

○ Under Construction HOT/Express Lanes
163 corridor-miles

● Existing HOV Lanes
722 corridor-miles

Source: Texas A&M Transportation Institute.

Figure 3: THE RANGE OF TOLLS FOR ROADS AND BRIDGES
Tolls Charged on Highways, Bridges, and Tunnels in the United States Vary Greatly

Road	Toll Rate for Passenger Cars with Electronic Pass (cents/mile)	Road	Toll Rate for Passenger Cars with Electronic Pass (cents/mile)
Indiana East–West Toll Road	3.0	Pennsylvania Turnpike	8.0
Dallas North Tollway	4.0	Blue Star Turnpike, NH	9.0
Kansas Turnpike	4.0	New Jersey Turnpike	11.5
New York State Thruway	4.5	Triangle Parkway, NC	15.0
Turner Turnpike, OK	4.5	State Highway 130, TX	15.0
Ohio Turnpike	4.5	Dulles Toll Road, VA	20.0*
Garden State Parkway, NJ	5.0	Intercounty Connector, MD	25.0*
Illinois Tollway	6.0	E-470 Express Toll Lanes, CO	26.0
Massachusetts Turnpike	6.0	Dulles Greenway, VA	34.0*
Florida Turnpike	6.5	Delaware Turnpike	36.0
Maine Turnpike	7.0	Chicago Skyway	45.0

* Peak Rate.
Source: ULI survey of toll authority websites, November 2012.

Bridge/Tunnel	Toll Rate for Passenger Cars with Electronic Pass
California	
San Francisco Golden Gate Bridge	$5.00*
Bay Area: Antioch Bridge, Benicia-Martinez Bridge, Carquinez Bridge	$5.00*
Dumbarton Bridge, Richmond-San Rafael Bridge, and San Mateo–Hayward Bridge	$5.00
San Francisco–Oakland Bay Bridge	$6.00
Delaware–New Jersey	
Delaware Memorial Bridge	$4.00*
Florida	
Sunshine Skyway	$1.25
Illinois–Indiana	
Wabash Memorial Bridge	$0.50
Louisiana	
Lake Pontchartrain Bridge	$3.00*
Massachuetts	
Ted Williams Tunnel	$3.50*
Maryland	
Baltimore Harbor Tunnel, Francis Scott Key Bridge, Fort Mchenry Tunnel	$3.00
Chesapeake Bay Bridge	$4.00*
New Jersey–Pennsylvania	
Delaware River Joint Toll Bridge Commission	$1.00
New York	
Marine Parkway–Gil Hodges Memorial Bridge and Cross Bay Veterans Memorial Bridge	$1.80
Bronx–Whitestone Bridge, Triborough (RFK) Bridge, Throgs Neck Bridge, Queens–Midtown Tunnel, Brooklyn-Battery Tunnel	$4.80
Henry Hudson Bridge	$2.20
Verrazano-Narrows Bridge	$9.60*
Lincoln Tunnel, Holland Tunnel, George Washington Bridge, Bayonne Bridge, Goethals Bridge, Outerbridge Crossing	$10.25*
Virginia	
Chesapeake Bay Bridge-Tunnel	$12.00**
Washington	
State Route 520, Seattle	$3.59 peak

* Toll charged for travel in one direction only.
** $5.00 return within 24 hours.
Source: ULI survey of bridge and toll authority websites, November 2012.

But land value is also influenced by decisions the public sector makes about how to pay for transportation investments.

Imagine two scenarios for paying for a new high-speed road from a central city to outlying areas, in a simplified world made up of a metropolis with a growing population and economy. In the "free road" scenario, the government uses general revenues to fund construction of this new high-speed road and allows users to drive on it for free. In the "toll road" scenario, the government uses tolling to finance road construction and charges a toll to use this new high-speed road. It's the same road, but the government uses two different approaches to pay for it.

Free Road and Land Use

In the short run, land served by the free road will be more desirable—more valuable—than land served by the toll road. But there is a limit to the desirability of land in the free road scenario, because the free road will experience congestion much sooner than the toll road. Once congestion sets in on the free road, the value of the land stagnates or even declines, because at peak demand the land served by the free road is no longer as close—in terms of travel time—to desirable locations.

The road managers, authorized by the government to look after the road, have only a limited range of responses to congestion on the free road. They can add capacity by building new lanes, but to do this means convincing government decision makers that an investment in new lanes is a higher priority than other uses of general tax revenue. The road managers can try to reduce demand for space on the road by encouraging carpooling or taking mass transit. They can even give incentives to carpoolers and bus users by designating a special lane just for them. And there are some traffic management technologies that can offer marginal relief.

But if the congestion is severe enough, these measures are only bandages—especially if the buses are still stuck in congestion—on the real problem, which is that the road managers seemed to have promised that this land would be, for example, 20 minutes from desirable locations and now it is 40 minutes from them. Even if new lanes are added, if the travel behavior and land development response is strong enough, congestion will eventually return.

In addition, the road managers have an incentive to protest additional land development near the highway. They will express worries about land development's "traffic impact." People regularly stuck in the congestion will join the road managers in protesting additional land development. In the very long run, if the road gets congested enough, the land uses begin to move—creating new desirable locations—so that they can be closer to each other, in terms of travel time, if not distance.

Toll Road and Land Use

In the toll road scenario, land development will be slower initially. People will take into consideration the cost of the tolls when making decisions about whether to move outward. At first, the road managers have an incentive to keep tolls low to attract users. The government may even find it advantageous to offer subsidies to land development in order to induce use of its toll road.

Once congestion sets in, however, it is a whole new ball game. The road managers can now vary the tolls to manage traffic levels, charging more for peak times of the day or for free-flowing lanes. To meet policy objectives, the road managers can set tolls to maximize revenue, speed, or the volume of vehicle flow. The road managers can even offer cross-subsidies, dropping toll rates during uncongested times.

Because the toll road managers gain revenue from congestion, they will be slower to respond with new capacity investments. They are also less likely to protest additional land development. (Since additional land development leads to higher tolls, once road users catch on to how the system works, they will still probably protest land development.) For those who can afford the tolls, a reliable, fast road is waiting for them.

One road, two different ways of paying for it, two very different land use futures.

From Free Roads to Tolled Roads

The American tradition of paying for roads with the revenue from taxes on motor fuels is like the "free road" scenario. Despite a long history within the transportation policy community of referring to taxes on motor fuels as "user fees," from the land use perspective, these taxes have functioned more like paying for transportation infrastructure from general revenue sources.

American taxes on motor fuels are, and have been historically, too low to be given much weight when land use decisions are made. Taxes on motor fuels, moreover, by their very nature are split between taxing fuel efficiency (or rather, inefficiency) and miles driven. As excise taxes collected from suppliers, they are also hidden. Few drivers understand how much in taxes they are paying per mile. Also, taxes on motor fuels are not affiliated with specific transportation investments. For example, even if you rarely drive on the Interstate Highway System or on other federal-aid highways, you still pay the federal tax every time you purchase a gallon of fuel.

Figure 4: REVENUE MECHANISMS AND THEIR RELATIONSHIP TO MOTOR VEHICLE TRAVEL

HIGH RELATIONSHIP TO DRIVING

VMT Taxes/Fees

Tolls   Taxes on Motor Fuels

Vehicle Registration Fees/Taxes

Property Taxes   Income Taxes

Sales Taxes

LOW RELATIONSHIP TO DRIVING

In essence, government policy—at both the federal and state levels—has kept the tax burden on passenger vehicle trips so low and so hidden that these taxes are functionally irrelevant to decisions about driving. When travelers choose whether to drive and to which destinations, variables such as travel time, parking costs, and fuel costs are much more important than taxes on motor fuels. Of course, the "entry fee" to this government-sponsored transportation network is access to a motor vehicle.

Although policy in the United States has been to keep taxes on driving extremely low, there are many who argue that charging too little has its own negative consequences. These critics point to a long list of costs associated with driving: air pollution, vehicle injuries and fatalities, dependence on foreign sources of oil, and the overdevelopment of land that could be reserved for agriculture or natural systems. If revenues raised from taxes on motor fuels covered these external costs of driving, estimates indicate that the tax rate would need to increase to as much as $2.00—or more—per gallon. Most developed countries go even further and have chosen to tax driving as a way to raise general revenues that fund more than their transportation systems. In Spain, France, Germany, and the United Kingdom, which assess both an excise tax and a value-added tax on motor fuels, taxes on gasoline range from $3.30 to $4.70 per gallon.

··

Figure 5: REVIEW OF TRANSPORTATION REVENUE MECHANISMS

Road user charge: A generic term that includes a range of mechanisms used to collect revenues from travelers according to their use of roadways. Although what is included in the category "road user charge" can vary, its broadest definition encompasses vehicle registration charges, vehicle weight charges, taxes on motor fuels, miles-traveled charges, and tolls. Road user charges may be taxes or fees.

TYPES OF ROAD USER CHARGES

Tax: A mechanism that governments use to raise revenue. Even if part or all of the revenue stream is dedicated to specific purposes, if the revenue source can be used for general purposes, the charge is still a tax.

Fee: A mechanism that governments use to raise revenue that can be used only for specific purposes and that is typically tied directly to the cost of providing the service for which the fee is charged. Many states have strict restrictions on the use of revenues generated via fees. These restrictions can have the effect of limiting the amount that can be charged via a fee.

Excise tax on motor fuels: A tax collected from suppliers of gasoline, diesel, and other fuels that power motor vehicles. The tax is charged per unit sold (i.e., per gallon) and in its current form does not vary according to the price of fuel. The federal government and all 50 states and the District of Columbia levy excise taxes on motor fuels. Some states also charge sales taxes on motor fuels, paid directly by the consumer.

Vehicle-miles traveled (VMT) tax: A tax collected from owners of motor vehicles and calculated per mile driven. In accordance with the legal context, a government may be able to vary the VMT tax according to region, time of day, road type, or other criteria established by the government.

Mileage-based user fee: A fee collected from owners of motor vehicles and calculated per mile driven. It could also be called a VMT fee. In accordance with the legal context, a government may be able to vary the per-mile fee according to region, time of day, road type, or other criteria established by the government. Because it is a fee, there would be restrictions on the use of the revenue based on the applicable legal context in the state.

Tolled roadway: A segment of roadway where the owner or operator charges a toll—a fee or tax—for access (e.g., crossing a bridge) or per mile of travel. Tolls may be set according to cost pricing, value pricing, or some combination, or tolling levels may be set according to other objectives, usually decided in the political arena.

TYPES OF PRICING FOR TOLLS

Cost pricing: A way of setting the price for transportation based on covering the costs of building, maintaining, and operating a facility or network. A flat charge to access a tolled roadway set at a level designed to pay back construction bonds over a number of years is an example of cost pricing. Cost pricing may also include indirect costs such as the *cost of motor vehicle crashes or air pollution.*

Value pricing: A way of setting the price for transportation based on what the traveler thinks it is worth and is willing to pay, rather than based on the cost to provide it. Value pricing can be used to set a price for free-flowing travel, but can also be used to set a price for other travel amenities such as truck-free (or car-free) lanes, express lanes that bypass local traffic, lanes for extreme speeds (above 75 mph), queue-jumping, and even pavement quality.

Congestion pricing: A type of value pricing that sets the price for road travel to reduce congestion or to maintain free-flowing conditions. Congestion pricing is usually variable: the price changes as demand rises and falls over the course of the day. Congestion pricing may also be dynamic: the price may change in real time as sensors monitor traffic levels. The most common use of congestion pricing in the United States is to provide free-flowing lanes on congested segments of urban highways. See managed lane.

TYPES OF MANAGED LANES

Managed lane: A lane of a roadway is called "managed" if it is proactively operated to achieve an objective, usually free-flowing conditions. According to the U.S. Federal Highway Administration, there are three main ways to manage a lane:

- use pricing;
- limit eligible vehicles (allowing, for example, only carpool vehicles and electric hybrids); and
- limit access to interchanges or intersections (e.g., traditional express lanes that bypass local exits).

The most common use of managed lanes in the United States is on congested segments of urban highways.

High-occupancy vehicle (HOV) lane: A lane designated for the exclusive use of vehicles meeting minimum occupancy standards, usually two but sometimes three persons per vehicle.

High-occupancy/toll (HOT) lane: A lane designated for high-occupancy vehicles that charges a toll to vehicles not meeting the minimum occupancy standard.

Federal transportation leaders in the United States are in the midst of rethinking a long history of anti–toll road policies. For decades, federal dollars could not be used to fund or otherwise support toll roads, and the federal government imposed penalties if state governments wanted to use tolling to reconstruct or expand roads built with federal aid. Only in recent years has federal policy begun to tentatively support tolling through small pilot programs allowing experiments with HOT lanes and through the Transportation Infrastructure Finance and Innovation Act (TIFIA) loan and loan guarantee program.

While the threat of federal penalties still dissuades states from attempting to convert existing federal-aid highways to toll roads, the most recent federal transportation law, MAP-21, opens the door a little wider for expanded tolling.

MAP-21, signed into law in July 2012, enables states to toll new capacity on the Interstate Highway System and other federal-aid highways. (States have always retained the authority to build and maintain toll roads constructed without federal funds.) The law also allows states to convert all existing high-occupancy vehicle (HOV) lanes to HOT lanes. Toll revenue must be used first for the facility itself. Only after the responsible public official "certifies" that the facility is being adequately maintained may the revenues be used for other transportation purposes. This certification must be done annually. Private sector partners, however, are allowed to earn a "reasonable return on investment."

MAP-21's endorsement of converting HOV lanes to HOT lanes builds on two pilot programs dating to the mid-2000s. The U.S. Department of Transportation (USDOT) Urban Partnership Program used discretionary grant funding and a competitive application process to fund congestion-relief projects that exemplified "the four Ts": tolling, transit, telecommuting, and technology. Among the projects funded are the following:

- Miami's I-95 express lanes (see case study, page 20);

- Minneapolis's I-35 West dynamically priced HOT lanes, including six new park-and-ride facilities and express bus service; and

- Seattle's State Route 520 variably priced toll bridge over Lake Washington, including bus service enhancements.

The USDOT's Congestion Reduction Demonstration Program funded two additional projects converting HOV lanes to HOT lanes:

- Los Angeles's I-10 and I-110 dynamically priced HOT lanes, including improvements for bus rapid transit routes; and

- Atlanta's I-85 dynamically priced HOT lanes, including bus transit improvements.

MAP-21 also dramatically expanded the TIFIA loan and loan guarantee program. What had been a small, experimental program (just $122 million annually) rises to $750 million in fiscal year 2013 and then to $1 billion in 2014. Toll revenue can be used to secure TIFIA support, and the majority of "letters of interest" submitted to the USDOT to date request support for toll road projects.

COURTESY OF ANDREW TUCKER

Bus riders descend from 46th Street to the I-35 West HOT lanes for express bus service. The HOT lanes project, pictured above and left, was funded under the USDOT's Urban Partnership Program.

Tolling, with and without value pricing, and taxes or fees charged per mile will influence land use decisions much more directly than taxes on motor fuels. Compared with taxing motor fuels, tolling (and charges per mile can be thought of as a toll on all roads) falls much more heavily on decisions people make about locations: where to live, where to work, where to shop, where to go to school, where to get health services, where to get supplies for a business, and so on. In effect, tolls are a charge paid by users of a specific piece of ground—a length of road—at a specific time. Another way to think of tolling is as a charge for accessing destinations.

Compared with motor fuel taxes, tolling has different psychological as well as economic impacts on driving and therefore land use. Tolls are not hidden. Moving to a system of tolls or charges per mile has the potential to raise awareness of the cost of driving and reveal the true value of specific transportation services such as high-speed expressways or free-flowing lanes. Theoretically, the economic impact combines with the psychological impact to induce an even stronger land use response.

Conclusion

Tolling and charging per mile—transportation revenue mechanisms expanding or under serious consideration in the United States—have a much stronger link to land use than taxes on motor fuels and other revenue mechanisms such as sales taxes and income taxes. Now is the time to include land use and the effects on land use decision making in research, policy discussions, and political debates around how revenue is raised from transportation.

To raise awareness of these coming changes and to frame the issues for further research and development, the ULI Infrastructure Initiative constructed and conducted an in-depth thought experiment for leaders in transportation and land use from across the United States. Through workshops and interviews, study participants explored and analyzed the interactions between land use and congestion pricing, tolling, and VMT taxes. Participants also discussed the potential for these transportation and land use interactions to have impacts on lower-income groups. The results of the ULI Infrastructure Initiative thought experiment are discussed in Part 2.

PART 2:
Land Use and Tolls, VMT Taxes, and Congestion Pricing

Concerned that there seemed to be little discussion going on in policy and implementation circles about how the renewed emphasis on tolling and the spread of value pricing will interact with land use in the United States, the ULI Infrastructure Initiative undertook a research effort into the land use impacts of tolling, VMT taxes, and congestion pricing. ULI convened and interviewed a group of transportation and land use experts and asked them to consider how these revenue mechanisms might affect metropolitan development and land use decision making over the next 20 years.

Study participants explored and refined five scenarios that illustrated the interactions between transportation revenue mechanisms and land use. Two scenarios probed different ways of implementing value pricing on managed lanes. Three scenarios examined the land use impacts of increasing taxes on motor fuels, charging for every mile driven, and converting the Interstate Highway System to toll roads. The ULI Infrastructure Initiative analyzed the discussions and produced qualitative descriptions of potential impacts.

....

Figure 6: QUICK SUMMARY OF THE THOUGHT EXPERIMENT SCENARIOS

TWO SCENARIOS: VALUE PRICING ON MANAGED LANES

1. Bus toll lanes   2. Optional toll lanes

THREE SCENARIOS: INCREASED TAX RATES AND TOLLS

3. Increased federal tax rates for motor fuels

4. Federal tax on all vehicle-miles traveled

5. Toll the Interstate Highway System

....

Study participants converged on a set of conclusions about the scenarios and their land use impacts.

General:

- The impacts of the scenarios' transportation revenue mechanisms on land use will be real, but modest, and will unfold over the long term.

- The cost of driving will become more important in decisions about land use and where to locate businesses, residences, and services.

- The impacts on land use will vary greatly by metropolitan region and will be influenced by the existing transportation network, including mass transit services, as well as by current land use patterns, local land use policies, and future regional economic trends.

- Value pricing on managed lanes, increased taxes on motor fuels, VMT taxes, and the spread of toll roads have the potential to interact with land use in ways that support growing market preferences for development in compact, mixed-use, walkable districts and neighborhoods.

- Absent careful coordination with land use policies and other transportation services, however, the potential to support the growing market preference for compact, walkable development is unlikely to be fully realized.

Value Pricing on Managed Lanes:

- Managed lanes incorporating value pricing offer potential benefits for land development.

- Crucial policy decisions about how value pricing is implemented—including policy decisions about toll levels, toll discounts for multi-occupant vehicles, coordination with transit service, and use of any excess revenue—will determine how broadly U.S. communities benefit from value pricing and managed lanes.

Expanded Tax/Toll Scenarios:

- Scenarios that increase the cost of driving provide incentives for development in compact, mixed-use, walkable nodes.

- Taxes on vehicle-miles traveled will affect more land use decisions and interact more strongly with land use than equivalent taxes on motor fuels or the expanded use of tolling.

- Scenarios that increase the cost of driving without also taking into consideration how to expand the transportation and land use benefits broadly among income groups have the potential to have a disproportionate impact on lower-income groups, especially lower-income groups dependent on driving.

The ULI Infrastructure Initiative convened discussions of the revenue mechanisms and possible implementation scenarios, including increased tax rates, for analysis purposes only. The inclusion of revenue mechanisms in this study does not imply endorsement.

Introduction to the ULI Thought Experiment

There is little empirical research—hard data—on the land use impacts of tolling, managed lanes, and VMT taxes. Shifts in American driving behavior, moreover, raise questions about whether it is appropriate to apply past research to present circumstances.

Research on American driving behavior going back decades once led to the conclusion that driving is economically inelastic: Americans drove more and more, regardless of rates charged for tolls, taxes, and fees. But American driving behavior is changing. In a phenomenon the *Economist* has dubbed "peak car," total driving in wealthy nations including the United States has plateaued and for certain age groups is declining. Early evidence also suggests that driving behavior may be becoming much more price sensitive than backward-looking data would suggest. It is time to reconsider the assumption that tolls, taxes, and fees on driving will have little impact on land use in the United States.

In addition, Americans have limited experience with implementing these revenue mechanisms, let alone analyzing how they interact with land use. Many regions in the United States are unfamiliar with toll roads. In those places with toll roads and bridges, toll prices have only recently started to rise to heights that might affect trip-making decisions. State departments of transportation have conducted only a handful of experiments with charging per mile. Most of these studies set out to test the technology, although they also may include some analysis of how people responded to being charged per mile. Value pricing to preserve free-flowing traffic is also very new. The expanded use of tolling and congestion pricing brings with it the ability to build new capacity and provide new amenities, but also new uncertainties for land development.

A SIMPLE MODEL

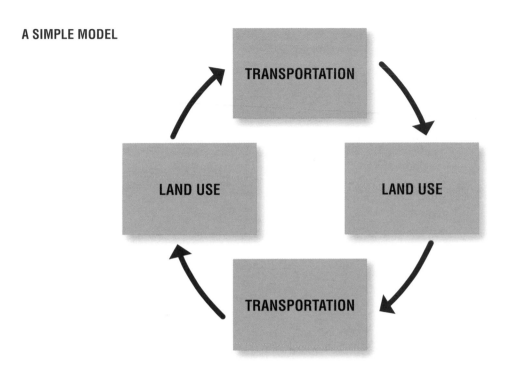

A MORE COMPLEX MODEL

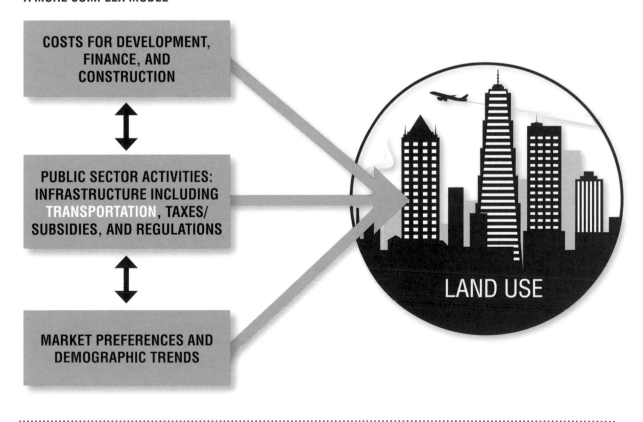

Confronted with the lack of hard data and the realization that even past empirical studies may have limited utility going forward, the ULI Infrastructure Initiative convened a group of experts to examine the land use impacts of congestion pricing, tolling, and taxing VMT. ULI held invitation-only workshops in San Francisco and Washington, D.C., and conducted interviews in June and July 2012, creating a conversation between transportation and land use experts. Transportation experts, including those with backgrounds in engineering, planning, economics, and policy and representing research institutes, state departments of transportation, and engineering firms, discussed the coming transportation changes and challenges with developers, investors, land use planners, and experts in land use policy and real estate economics.

Focusing on changes to land use, the experts explored and analyzed three scenarios imagining the adoption of one of three transportation revenue mechanisms: increased taxes on motor fuels, tolling the Interstate Highway System, and taxes on vehicle-miles traveled. Study participants also discussed using value pricing to manage congestion on expressways and developed two separate scenarios that reflect different policy choices and their outcomes. Collectively, the experts indicated the types of potential impacts on land use and commented on the impacts' likely magnitude. They also considered how the scenarios and their land use impacts would affect lower-income groups.

From these discussions, the ULI Infrastructure Initiative developed qualitative descriptions of the interactions between the revenue mechanisms and land development, which are outlined in the sections below. For purposes of clarity, the presentation simplifies land use decisions, focusing only on the interactions between the transportation mechanism itself and land use. In real-world situations, a multitude of factors affects each land use decision. In any given land use decision, factors other than transportation are likely to be equally important as or more important than the cost of transportation taxes, travel time to destinations, or the availability of transportation alternatives (see Figure 7).

Mechanisms Will Have Real, but Modest, Impacts

A consensus arose among workshop participants and interviewees that the transportation revenue mechanisms under consideration will interact with land use. Robust discussions about the direction of change for specific land uses (i.e., attract versus repel, encourage versus discourage, accelerate versus inhibit) also occurred. Study participants, however, expressed much more uncertainty and much less agreement about the magnitude of the impacts. Behavioral research to date is not complete enough to lead to confident predictions. The magnitude of the impacts, moreover, will depend on local factors, including the availability of transportation alternatives and substitute destinations. People will also respond differently—make different land use choices—depending on other variables, including income levels, stage of life, and type of business or employment.

Workshop participants and interviewees predicted that—as a general proposition—tolling, congestion pricing, and taxing VMT will have real, but modest, impacts on metropolitan land use. In other words, if the use of tolling, congestion pricing, and/or VMT taxes is expanded, 20 years from now U.S. metropolitan landscapes will look different. The changes, however, are more likely to be subtle, rather than dramatic. If dramatic changes do occur, they are the most likely to take place along specific corridors or within areas of concentrated development.

Figure 8: SUMMARY OF THE FIVE SCENARIOS AND THEIR IMPACTS

Land Use Impacts of Transportation Revenue Mechanisms: Real, but Modest

TWO SCENARIOS: VALUE PRICING ON MANAGED LANES

1. Bus toll lanes

- The network of lanes is part of the rapid transit system.

- Highest priorities are keeping transit users moving and providing a reliable travel time to all paying users.

- Excess revenue is used for transit service and to improve infrastructure at destinations.

Land use
Help meet demand for and foster development in compact, mixed-use, walkable nodes.

2. Optional toll lanes

- Toll lanes bypass existing congested bottlenecks.

- Highest priority may be speed, revenue, or reliability and is decided on a case-by-case basis.

- Excess revenue is used for highways elsewhere in the region and state.

Land use
Likely to support sprawl.

THREE SCENARIOS: INCREASED TAX RATES AND TOLLS

Revenue mechanism	Land use
Baseline: No expansion of revenue mechanisms	Existing conditions continue, providing no encouragement of market trends for development in compact, mixed-use, walkable nodes.
3. Increased federal tax rates for motor fuels	Encourage current market trends for compact, walkable development.
4. Federal tax on all vehicle-miles traveled	Accelerate development in compact, walkable nodes
5. Toll the Interstate Highway System	Interactions with land use depend on local and metropolitan factors.

Land Use Impacts Are Long-Term Impacts

ULI asked study participants to consider impacts five years and 20 years after the implementation of each transportation revenue mechanism, because shifts in land use reveal themselves over the long term.

When faced with increases in the cost of driving, people adapt in the near term by making different choices in transportation and then in other areas of discretionary spending. For example, a two-car family may start driving the fuel-sipper more often than the gas-guzzler, or families may shift resources from the dining-out budget to cover the increased cost of commuting.

I-95 EXPRESS LANES

BROWARD AND MIAMI-DADE COUNTIES, FLORIDA

Variable pricing	Free to HOVs	Coordinated with bus service	Heavy trucks (more than two axles)	Electronic toll collection	Peak charge per mile (approximate for passenger vehicles)
Dynamic pricing according to demand	HOV-3+	Yes	No	Yes	$0.41 southbound and $0.62 northbound average peak with $0.95 maximum

Miami's first HOT lanes opened in December 2008. The Florida Department of Transportation (FDOT) took the lead on this joint highway/transit project that has significantly improved the travel experience for commuters, business travelers, and transit users.

FDOT took an "all hands on deck" approach to speed up vehicle movement along 7.3 miles of I-95 north of downtown Miami. On a stretch of highway where even the carpool lane regularly slowed to a crawl, FDOT narrowed travel lanes and used a bit of the shoulder to add a second HOV lane. In addition to an increase in the HOV occupancy minimum from two to three, the project added a toll option giving all passenger vehicles and light trucks the opportunity to travel in the HOV lanes. Dynamic pricing manages demand in the tolled lanes to keep traffic moving above 45 miles per hour, a target that the express lanes have been able to

maintain 95 percent of the time. They also introduced ramp metering to interchanges.

> *"The goal is to keep traffic in the express lanes moving at a minimum speed of 45 mph while maximizing person throughput of the entire facility."*
>
> —FDOT, *95 Express Annual Report*, February 2012

Improvements to transit service and park-and-ride facilities were coordinated with the highway investments. Four new express bus routes, departing as frequently as every 15 minutes during peak commute times, connect communities in the corridor to downtown Miami. There is also weekend bus service. Bus ridership has increased 145 percent since the express lanes opened;

Land use impacts appear incrementally, as certain locations become more (or less) attractive. In a process that plays out over decades, opportunities begin to appear for expanding development in favored locations. New building types and site development innovations, better adapted to the new transportation pricing regime, eventually take hold and spread. But the obsolescence of less favored locations and outdated building types can be painful for their host communities.

Study participants analyzed two types of long-term land use impacts:

Changes in trip making affect land use. For this kind of impact, tax/toll hikes or new value-priced amenities lead to adjustments in the transportation arena, some of which lead to people making different choices about destinations. These land use impacts are related to changes in trip-making patterns, which are most likely to affect patterns for patronizing shopping and service locations. These changes may happen quickly or slowly, and they have the potential to permanently alter land use patterns.

during peak periods, express bus riders account for 18 percent of express lane travelers. In FDOT surveys of transit users, 53 percent of new riders said the express lanes influenced their decision to take transit and 38 percent reported they formerly drove.

The speed and reliability of the express lanes have also attracted business users who travel the lanes throughout the day. Business travel represents 57 percent of managed-lane use, with these noncommuters coming from a diverse set of industries including health care, legal services, management, and construction.

Phase II of the I-95 express lanes project, scheduled to open in 2014, will expand the HOT lanes nearly 14 miles north into Fort Lauderdale, Florida.

Funding for the highway and transit capital improvements ($132 million for Phase I; $106.1 million for Phase II) came from a combination of federal, state, and local sources. Toll revenues are designated to facilities operations first, with 20 percent of revenue going to express bus operations. Revenues exceeded projections by 15 percent in the first full year of operations.

The success of the I-95 express lanes is building enthusiasm for expanding the HOT lane network throughout the region. Construction of a reversible HOT lane with bus service is underway on I-595, and plans call for expanding the toll lanes, express bus service, and park-and-ride facilities to other expressways.

Express bus riders on I-95 experience a congestion-free trip.

General increases in the cost of driving, for example, encourage behavior changes such as:

- Walking to neighborhood restaurants instead of driving to dine across town.
- Shopping online instead of making trips to the regional mall.
- Substituting conference calls for in-person meetings.
- Consolidating retail trips—for example, making one trip instead of three to the grocery store in a week. This, in turn, reduces the opportunity to impulse-shop.

Amenities or increases in the cost of driving that are route-specific encourage, for example:

- Choosing to reroute a commute home to avoid a toll road, and as a consequence patronizing a different grocery store and gas station along the new route.

- Choosing to reroute a commute home to take advantage of a toll road, giving up the trip-chaining along "big-box row" and instead running errands closer to home during the evening and on weekends.
- Taking advantage of the reliability provided by new value-priced managed lanes to use a daycare close to home instead of close to work.

The cost of driving influences major decisions about locations. Here, land use changes begin to accumulate as people make major decisions involving locations such as buying a new house, expanding a business, or looking for a new job. The new transportation pricing mechanisms will be taken into consideration, and it can be expected that people will give more weight to the costs of accessing different locations, the risks of sudden price changes, and the transportation alternatives available. Land developers and investors will explore whether including these same factors in their business models gives them a competitive advantage.

Metropolitan Development Trends

ULI focused the study of the land use impacts of transportation revenue mechanisms on metropolitan areas (with populations exceeding 50,000) and metropolitan development patterns for housing, office, retail, and industrial uses. Tolling, congestion pricing, and taxing VMT will also interact with rural land uses and affect patterns of intercity travel, including mode choice, in ways that have implications for airports, rail and bus stations, and intermodal freight facilities. The impacts on rural areas and implications for intercity travel, while not the focus of this study, deserve additional analysis.

Land use experts participating in the study noted that analysis of the land use impacts of transportation revenue mechanisms builds on an understanding of existing land use patterns and future trends for land development. Tolling, taxing VMT, and congestion pricing are most likely to have the biggest impact on new development because developers and investors can respond directly to the amenities and price signals that the transportation revenue mechanisms provide. Therefore, understanding the trend for new development is key to understanding how transportation revenue mechanisms will affect land use.

Figure 9: TRENDS FOR FOUR TYPES OF METROPOLITAN DEVELOPMENT

Metropolitan Development Type	Today's Generalized Trend*
New development in compact, mixed-use, walkable nodes	Growing investment attractiveness
New development in sprawling, single-use areas	Stable or declining investment attractiveness
Existing development in compact, mixed-use, walkable nodes	Growing or stable investment attractiveness
Existing development in sprawling, single-use areas	Stable or declining investment attractiveness

* These trends are very generalized and should not be taken to represent the future prospects of any specific development site or proposal.

Most land use experts consulted for this study agreed that land use preferences in the United States are shifting. They foresee the largest increases in growth for new development located in compact, mixed-use, walkable nodes or activity centers. The sprawling, single-use districts for housing, office, and retail uses that dominated development over the past 50 years won't disappear, of course. But going forward, more new development will be built in compact, mixed-use, walkable nodes than past trends would indicate, and less new development will be built in sprawling, single-use districts.

ULI found four broad types of metropolitan development potentially affected by tolling, taxing VMT, and congestion pricing. Figure 9 summarizes the four and indicates today's development trends.

Two Scenarios: Value Pricing on Managed Lanes

For the value-pricing scenarios, ULI asked study participants to analyze how a network of reliable, high-speed expressway lanes managed with the use of value pricing will interact with land use and affect metropolitan areas. For study purposes, it was assumed that the value-priced lanes were either converted from existing lanes (such as an HOV to HOT lane conversion) or added as tolled lanes to the existing "free" expressway network, so that free lanes and pay lanes run parallel to each other throughout the network.

Study participants discussed the various ways in which managed lanes can operate, including the amenities that value pricing may offer to nearby land uses and how policy decisions about the operation of managed lanes interact with land use. Out of these discussions arose two very different alternative futures, which ULI packaged into the two scenarios presented below. The two managed-lane scenarios—in their approach to planning, operation, management, and use of revenues—are extremes illustrating opposite ends of a spectrum.

Value-Priced Lanes Offer Amenities for Development

Using pricing to create a network of free-flowing, reliable lanes offering expressway speeds 24 hours a day makes experiencing highway congestion a choice. Instead of congestion being something that just happens to a road user, drivers choose whether to sit in traffic or to pay to speed past it.

From a land use perspective, a network of value-priced expressway lanes offers two types of amenity values: speed/time-savings and reliability. Frequent users are more likely to be valuing the speed/time-savings and will be attracted to locations near managed lanes because, for example, managed lanes

- allow their business to make more service calls in a day; and
- allow them to live further away from potential job locations and still make the commute in less than 30 minutes.

Occasional users are more likely to be valuing the reliability. The existence of the managed lanes provides peace of mind for those times when a trip really needs to go smoothly. For both those willing—and able—to pay for frequent use and those only willing—or able—to pay for occasional use, the network of managed lanes increases mobility.

All other factors being equal, locations with easy access to the value-priced lane network will have a competitive advantage. As people make decisions about home, work, and business locations, areas accessible to managed lanes will attract those

Figure 10: THE RELATIONSHIPS AMONG SPEED, FLOW, AND PRICE

For value-priced lanes, meeting targets for speed, reliability, and revenues requires setting toll rates to achieve desired vehicle flows.

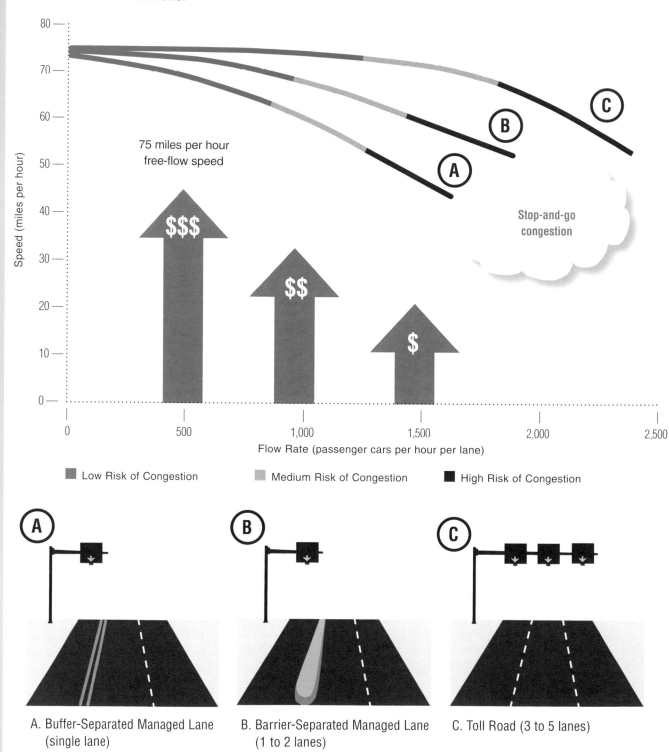

A. Buffer-Separated Managed Lane (single lane)

B. Barrier-Separated Managed Lane (1 to 2 lanes)

C. Toll Road (3 to 5 lanes)

Source: CH2M HILL, based on Transportation Research Board, Highway Capacity Manual 2010, and National Cooperative Highway Research Program 3-96.

who value time and reliability more highly. Because the free lanes and the priced lanes run adjacent to each other, locations near value-priced lanes offer less risk that residents, workers, and tenants will feel "priced out" of the road, if regional or personal economic circumstances don't go in their favor. Drivers can always take the free lanes. In addition, the network of value-priced lanes makes it easier to choose car-dependent land use patterns because it removes the aggravation of being forced to crawl along in bumper-to-bumper highway traffic.

Policy Decisions about Managed Lanes Influence Land Use

Using pricing to manage the flow of traffic requires a series of policy decisions to determine who accesses the facility and at what price. The outcome of these policy decisions will have much to do with how the road interacts with land use, and land use planners and developers need to understand the implications of the operational characteristics of the managed lanes as well as the physical characteristics of the infrastructure investment.

For a typical American expressway or freeway, there is a direct relationship between the number of vehicles in a lane and on the road and the speed at which traffic flows. For a value-priced lane, the road managers use pricing to control how many vehicles "choose" to enter a lane. The road managers set the price so that the lane, under normal conditions, never crosses the threshold where traffic flow quickly falls apart, creating the stop-and-go conditions that most people regard as congestion. But even under free-flowing conditions, fewer vehicles (via higher base tolls) accessing value-priced lanes allow traffic to fly at 75 miles per hour; more vehicles (via lower base tolls) slow traffic down to 55 miles per hour. And difference between the number of passenger vehicles "allowed" so that the lane operates at 75 miles per hour versus 55 miles per hour can be significant. Whether a 30-mile, value-priced commute takes 24 minutes or 33 minutes is a policy decision rooted in how the road is managed.

There are also options for the types of vehicles—and at what discount—to let into value-priced lanes. Is the objective to offer free-flowing traffic to the maximum number of vehicles or the maximum number of people? Should buses and vanpools get free entry? What about carpools? With two or three occupants? If multi-occupant vehicles are not allowed in for free, how much should the discount be? And as traffic increases in the managed lane, which should be sacrificed first: speed, single-occupant vehicles via escalating tolls, or discounts for multiple-occupant vehicles? Because all of these choices are policy decisions, the "right" mix of answers for a region or a managed-lane project will be decided in the political arena.

Just as important, the use of the revenue generated by the tolls is also a policy decision. With value pricing, the revenue generated by toll lanes bears little relationship to the cost of providing the facility. The amount of revenue generated is determined by the operational characteristics chosen and the amounts that users are willing to pay for congestion-free trips. The facility may generate revenue that barely covers the cost of toll collection, requiring the use of other revenue sources for capital investment and periodic maintenance. At the other extreme, the facility may generate revenue far in excess of roadway needs. How to use any excess revenue is a policy decision.

SAN DIEGO, CALIFORNIA

Variable pricing	Free to HOVs	Coordinated with bus service	Heavy trucks (more than two axles)	Electronic toll collection	Peak charge per mile (approximate for passenger vehicles)
Dynamic pricing according to demand	HOV-2+	Yes	No	Yes	$0.20 average peak with $0.40 maximum

COURTESY OF THE SAN DIEGO ASSOCIATION OF GOVERNMENTS (SANDAG)

San Diego's I-15 corridor is a national leader in new toll technologies and in integrating toll lanes and bus service. Its carpool lanes converted to HOT lanes back in 1998, and it was the first to experiment with dynamic pricing to manage demand and keep traffic flowing. Premium express bus service connects the corridor's job centers, taking advantage of the reliable travel times. Toll revenue, split 55 percent for facility operations and 45 percent for bus service, consistently exceeds the revenues needed to cover operating costs.

Now, the corridor is in the midst of another significant expansion to both the highway and the transit service.

In January 2012, the highway part of the most recent evolution of the HOT lane system opened: 20 miles of I-15 express lanes consisting of four reversible lanes running down the center of I-15 from Escondido, California, to San Diego. (The old

In the I-15 corridor, premium express bus service connects to job centers and downtown. By 2014, infrastructure upgrades for a complete bus rapid transit network will be completed.

Scenario 1: Bus Toll Lanes Connect Compact, Mixed-Use, Walkable Nodes

In this scenario, managed lanes are planned as part of a network that connects existing and expanding compact, mixed-use, walkable nodes such as downtown or suburban activity centers. The managed lanes are developed strategically to fill in gaps in the high-capacity transit network and to strengthen connections to places with the potential to be strong transit destinations.

Buses and registered vanpools have priority and get free access to the managed lanes; all other vehicles are charged the same toll. Speed is managed to maintain bus performance, and infrastructure for mass transit is built into the network from its initial stages. The infrastructure includes park-and-ride lots at less intensely developed nodes and special facilities for embarking and disembarking buses that minimize user time. Toll and fare revenue is used to pay debt, maintain the facility, and operate the transit service and park-and-ride lots. Any excess revenue is used

HOT lanes provided two reversible travel lanes for eight miles.) When the transit component is completed in 2014, the corridor will sport a new bus rapid transit (BRT) network, the first in the region. New transit stations and upgrades to existing transit stations include revamped shelters, increased parking, and signs with real-time transit travel information. Direct-access ramps to the express lanes will further reduce bus travel times.

Peak service on the new BRT routes will be as frequent as every ten minutes, with weekend and late-night service available on some routes. One BRT route will operate like a rail line, moving up and down the corridor stopping at each station and in downtown San Diego. Another BRT route will connect to the campus of the University of California at San Diego. The BRT system is expected to boost transit ridership in the corridor by 41 percent.

The cost of expanding the express lanes and developing the new transit infrastructure totaled $1.3 billion. State and local funding sources, including a half-cent sales tax approved by San Diego region voters in 2004, provided most of the capital investment.

COURTESY OF THE SAN DIEGO ASSOCIATION OF GOVERNMENTS (SANDAG)

Toll revenue from the I-15 express lanes is split between facility operations (55 percent) and bus service (45 percent).

Even before this latest expansion of the I-15 corridor's express lanes and transit service, the HOT lanes were recognized as an important factor to those choosing to live in the region. A 2006 survey conducted by San Diego State University found that 27 percent of new residents rated the I-15 express lanes as very high or high in their housing location decision; 20 percent rated the toll option as very high or high in their housing decision.

in the nodes to support transit service, pedestrian/bicycle infrastructure, and other infrastructure that facilitates compact, mixed-use, walkable development patterns.

Impact on land use: bus toll lanes help foster and meet demand for compact, mixed-use development. This scenario provides amenity values for development in the nodes via three mechanisms. For residents and workers who use cars, there are the time savings and reliability—offered for a price—to areas of concentrated development. Residents and workers who use transit benefit from the enhanced regional mobility. The policy decision to use excess revenues from tolls and fares to pay for infrastructure that supports the economic vitality of the corridor and the nodes enables users to contribute to enhancing their experience in the corridor and at their destinations.

When transportation alternatives (e.g., pay lanes, free lanes, and transit service) are provided, a broad swath of the population benefits. Also, transportation costs

DENVER–BOULDER CORRIDOR, COLORADO

Variable pricing	Free to HOVs	Coordinated with bus service	Heavy trucks (more than two axles)	Electronic toll collection	Peak charge per mile (approximate for passenger vehicles)
Time of day	HOV-2+	Yes	No	Yes	To be decided

High-quality bus service welcomes riders traveling the 28 miles along U.S. 36 connecting Denver, Colorado's largest city, and Boulder, home to the University of Colorado. And transit service is about to get better, as the region upgrades the corridor with the infrastructure for bus rapid transit and develops toll lanes that will be managed to

The RTD bus stop and pedestrian bridge across U.S. 36 are an integral part of the Arista development in Broomfield, Colorado. Managed lanes will offer bus riders a congestion-free trip.

COURTESY OF ARISTA, BROOMFIELD, COLORADO

COURTESY OF ARISTA, BROOMFIELD, COLORADO

keep the buses moving at speeds above 50 miles per hour. The corridor communities are already embracing the new bus/toll infrastructure. The city of Broomfield, Colorado, for instance, is focusing new development at Arista, a bus-oriented node.

The four-lane highway opened as a toll road in 1951, but with repayment of the construction bonds, tolling stopped in 1968. Its original inter-city travel function—with only one interchange between Denver and Boulder—changed with sub-urban expansion. Today, ten interchanges access adjacent communities.

The U.S. 36 BRT project is managed by a partnership between the Regional Transportation District (RTD) and the Colorado Department of Transportation (CDOT) and is part of a larger effort to expand regional transit service including the light-rail and commuter-rail network. The RTD began the BRT project by upgrading the existing transit infrastructure along U.S. 36. Infrastructure improvements included moving the park-and-ride lots and transit stops immediately adjacent to the highway corridor, adding new bus-only slip ramps and queue-jump lanes, and constructing pedestrian bridges.

Completed in 2010 and 2011, these improvements have already saved bus riders between Denver and Boulder up to 15 minutes each way. Buses heading for the corridor depart from Denver's Union Station as frequently as every five minutes, and are permitted to drive in the highway shoulders when travel speeds drop below 35 miles per hour.

The full BRT plan will come to fruition with completion of the express lanes now under construction. CDOT plans to construct an additional single, buffer-separated toll lane in each direction. These express lanes are intended to make traveling between Denver and Boulder on the express bus (no extra stops) up to 24 minutes faster than the trip by car in the general-purpose—"free"—lanes. Even the bus routes with corridor stops are expected to be up to 17 minutes faster than travel by car in the general-purpose lanes.

The completed transit improvements cost $23.3 million, and were funded as part of a voter-approved regional sales tax increase dedicated to transit. The $425 million estimated cost of the express lanes is being funded by a combination of mostly local and state sources. The toll revenue is pledged to secure a $54 million federal TIFIA loan.

Broomfield (population 57,400), a suburban community located about halfway between Denver and Boulder, provides an example of a community coordinating BRT, value-priced lanes, and suburban mixed-use development. The old park-and-ride lot required buses to exit U.S. 36 and drive a half mile to pick up passengers. Buses now meet passengers in the highway corridor, via slip ramps and a pedestrian bridge, shaving six to eight minutes off the commute to Denver. Park-and-ride parking is integrated into Arista's mixed-use development. Of the 1,500 spaces, 560 are shared with the 1stBank Center, a 6,000-seat entertainment venue. Within an easy walk of Arista's U.S. 36 bus stop, housing options include live/work lofts, townhouses, and two apartment communities. Arista Place, a planned main street district, includes 158,000 square feet of Class A office space over street-level retail uses.

Buses now meet passengers in the highway corridor, via slip ramps and a pedestrian bridge, shaving six to eight minutes off the commute to Denver.

alone are not likely to drive residents and businesses out of the compact, mixed-use nodes along the corridor. If development in the nodes is not constrained, but is allowed to intensify, fewer residents and businesses will be "priced out" of living and working in them. If development in the nodes is constrained, the availability of transit service may not be enough to balance out the land value increases in the nodes, and affordability issues may be exacerbated for lower-income individuals.

Scenario 2: Optional Toll Lanes Access Sprawl

In this scenario, value-priced lanes are developed incrementally to offer a pay option to bypass congestion at known bottlenecks in the expressway network. Every passenger vehicle pays the same toll. When setting the toll and speed for the managed lanes, the road managers respond to political pressure from users. In some places, the base toll is set high and the cars fly; in other places, the base toll is set low and offers a trip slightly faster than that offered by the free lanes.

Mass transit providers can choose to use the managed lanes, but few of the disjointed segments run along important transit service routes or serve high-use transit destinations. There is no special infrastructure to facilitate bus or vanpool service. Toll revenue is used first to maintain the managed lanes and pay off debt. Maintenance of the adjacent free lanes has second priority. Any excess revenue goes into the state highway budget to be redistributed to high-priority highway projects elsewhere in the state.

Impact on land use: managed lanes provide amenities that serve car-dependent land development, but offer little else. This scenario would, over time, attract land uses inhabited by businesses and residents with a high value on—and who can pay for—speed and reliability. Because of the availability of the free lanes, transportation costs alone will not drive out land uses inhabited by those with a lower value on—or who have less ability to pay for—speed and reliability on the highway system. If developable land is constrained, however, businesses and residents with a higher value on speed and reliability may outbid those with a lower value on speed and reliability when buying or leasing homes or space for businesses.

Fear of this scenario is behind the pejorative nickname "Lexus lanes" that opponents have regularly tossed into political debates over value-priced lanes. Too few constituents see how they would benefit from value pricing, both as drivers and as people who access destinations. They fear, instead, that only a small, select segment of the population would take advantage of the managed lanes and that the managed lanes will not do much to expand transportation alternatives or meet growing demand for desired land use patterns and lifestyles.

Value-Priced Lanes on Arterial Roads

Up to this point, the discussion has assumed that managed lanes are available only on high-speed expressways. Technology makes it possible, however, to imagine using value pricing to manage traffic on a variety of roadway types, including heavily trafficked arterials with traffic signals. Tolling could be used to separate local and through-traffic or to allow queue-jumping at intersections. Participants in the ULI workshops and interviews only touched on the land use impacts of value pricing on arterial lanes. Many of these arterials already host combinations of automobile, truck, mass transit, pedestrian, and bicycle traffic. The arterials also provide local access to adjacent land development as well as facilitate regional connectivity. At the very least, adding a value-pricing component raises additional urban design challenges. Other land use impacts—both positive and negative—have not been explored, but could be significant.

Three Scenarios: Increased Tax Rates and Tolls

ULI also asked participants to consider scenarios in which the rates charged via transportation revenue mechanisms were significantly higher than today's taxes and tolls. Tax and toll levels were set to magnify the interaction between the cost of driving and land use, and do not reflect an endorsement, on the part of ULI or study participants, of higher taxation levels.

Because existing taxes on driving in the United States are low (and hidden), exploring small increases in the existing and proposed tax/fee mechanisms made it difficult to get a solid read on the impacts on land use. Study participants, therefore, were asked to consider tax rates that are much higher than likely to happen within the study's 20-year time frame.

In order to isolate the interaction between the revenue mechanism and land use, ULI also asked participants to consider several other assumptions:

- The tax increase is adopted at the federal level, so it applies to all metropolitan regions. No region gets a competitive advantage from adopting, or not adopting, increased taxes on driving.

- The tax increase is part of a larger reform of the federal tax code so that most Americans' total tax burden remains roughly the same.

- Revenues from transportation revenue mechanisms are placed in the general fund, and are no longer segregated into a highway trust fund.

- The Obama administration's recently adopted vehicle fuel efficiency targets are achieved.

The tax rates explored by the study participants are summarized in the table below. While these tax rates represent a dramatic increase in the current federal tax, they are still much lower than European taxes on driving. They would bring the United States roughly in line with Canadian taxes on driving.

In the real world, increased taxes and user fees that affect driving are likely to be much lower than those considered in the scenarios and are also likely to originate at the state level, not the federal level. On the other hand, toll rates for bridges and roads are already climbing across the United States, with the Chicago Skyway reaching 46 cents per mile. At the low end, toll rates come in at five to seven cents per mile, approximating the scenario's VMT tax rates. Current proposals for introducing VMT taxes typically start at two cents per mile for state government. Add in a penny at the federal level and a half cent at the metropolitan level, and a total VMT tax rate reaching nearly half the level examined in the scenario is not unforeseeable.

The land use impacts described below are extremes that policy makers, planners, and developers can learn from, but they will need to adapt the broad frameworks to specific transportation proposals and land use/transportation coordination efforts in their region.

Figure 11: TAX AND TOLL RATES USED IN THE THOUGHT EXPERIMENT

Increased Tax/Toll Rates for 2014–2034	Gasoline or Passenger Vehicles	Diesel or Trucks
Baseline: Current rates for federal taxes on motor fuels	$0.184/gallon	$0.244/gallon
Scenario 3: Increased taxes on motor fuels	$1.50/gallon	$1.75/gallon
Scenario 4: Federal tax on VMT	$0.073/mile	$0.282/mile
Scenario 5: Toll the Interstate Highway System	$0.318/mile	$0.724/mile

Baseline: No Expansion of Transportation Revenue Mechanisms

Federal and state taxes on motor fuels stay at current levels and states do not expand the use of tolling, including for managed lanes. This scenario projects the past into the future. Transportation investments above and beyond what current transportation revenue streams can pay for will have to compete with other needs for taxpayer dollars. Most states and metropolitan regions will probably still make the basic transportation investments needed to stay competitive, but the source of revenue will be income, property, or sales taxes, supplemented by specialized revenue sources such as impact fees, transaction taxes, and value-capture mechanisms. For individual vehicle users, the cost of driving remains low and highway congestion will still be a universal experience in large metropolitan areas.

Impact on land use: taxes on driving provide no encouragement to current market trends. Transportation taxes, fees, and tolls continue to be largely irrelevant (in the case of tolls, as relevant as they are now) when people make land use decisions. Existing trends in land use continue, with no additional signals to users and to the market from transportation revenue mechanisms.

Scenario 3: Increased Taxes on Motor Fuels

Among workshop and interview participants, the weight of opinion indicated that even a significant increase in taxes on motor fuels would have little impact on land use patterns in the United States. They argued that the increased tax rate on motor fuels would accelerate the adoption of fuel-efficient or alternative-fuel (e.g., electric) vehicles before it would affect land use. As an excise tax, a higher motor fuel tax is still hidden and not directly connected to each trip and destination decision.

Impact on land use: encourage current market trends for development. Those who thought the increased tax on motor fuels would influence land use decisions thought it would encourage development in compact, mixed-use, walkable nodes and give a competitive advantage to development and land in or adjacent to such nodes.

Scenario 4: Federal Tax on VMT

This tax, sometimes called a mileage-based user fee, charges road users for each mile of driving. The tax applies to driving on all roads from local streets to the Interstate Highway System. In the study scenario, it is a tax and not a user fee because it is a revenue source for the general fund. The scenario increases the cost of driving, but it also increases awareness of the cost of driving. Schemes for VMT taxes propose some form of metering in the vehicle, and drivers receive regular bills, like utility bills, that tally total driving. To allow drivers to opt out of the tracking required for billing, some proposals allow users to prepay anonymously, like for disposable cell phones, or allow users to buy "unlimited miles" plans.

As a VMT tax, the tax falls completely on decisions made about the location of destinations, i.e., decisions about where to live, work, and play. The only way to avoid or minimize the tax is to drive fewer miles. This is why a tax (or fee) on miles driven will have a much stronger impact on land use than an equivalent tax on motor fuels. The scenario also assumes that the VMT tax would be a flat tax, where every mile driven is taxed at the same rate.

Impact on land use: accelerate development in compact, mixed-use, walkable nodes. Study participants agreed that the VMT tax scenario would provide momentum to the growing market preference for compact development. Urban core areas, suburban activity centers, and small metropolitan areas would benefit over sprawling suburban and exurban areas within megaregions. The VMT tax scenario would also accelerate the expanding preference for walkable districts and neighborhoods and increase demand for mass transit and ride-share services.

Study participants drew particular attention to the interaction between housing and job location decisions. They identified two broad groups who would respond to the VMT tax scenario in different ways:

- For skilled workers, jobs are typically clustered in downtowns or other activity centers. They therefore gravitate toward housing locations that are accessible to skilled-job clusters. They may even find a job first, and then look for housing. The impact of the VMT tax scenario would be to shrink the size of the area considered desirable for housing.

- For unskilled workers, jobs are dispersed and ties to home communities are typically stronger. The impact of the VMT tax scenario, therefore, would be to shrink the size of the area that they are willing to consider for jobs.

Study participants also discussed the impacts of the VMT tax scenario on industrial, office, and retail land uses. In all three, access to unskilled workers was the most significant impact identified.

Impact on industrial uses. The location of industrial uses is not likely to be affected by the VMT tax scenario because the need for large expanses of land, the attraction of fixed locations such as ports, and the other benefits of traditional industrial sites are still likely to dominate industrial location decisions. The fixed location of industrial uses, however, has the potential to magnify the issue of access to skilled and unskilled workers. Depending on local circumstances, industrial employers may find they need to subsidize employee transportation, or offer services to help provide housing and community facilities such as schools or shopping for workers closer to the industrial plant. Or they may need to offer higher wages to compensate for the higher transportation costs.

Impact on office uses. For the most part, whether office uses are concentrated in mixed-use nodes or dispersed will depend on other factors. For office market segments where developments within and outside of compact, mixed-use nodes compete with each other, the VMT tax scenario will give a competitive advantage to office development within the node. For office tenants, the VMT tax scenario could raise issues about access to unskilled workers in the same way that it does for industrial employers. Since office locations are more varied than industrial sites, difficulties in attracting unskilled office workers will be very location-specific.

Impact on retail uses. Retail uses are affected by the quality of their access to three different groups: customers, workers, and the transporters of goods. Study participants foresaw little impact on retail land uses from the perspective of goods movement. The vast majority of the impacts on retail uses will be tied to customer access and ability to attract workers.

INTERCOUNTY CONNECTOR

MONTGOMERY AND PRINCE GEORGE'S COUNTIES, MARYLAND

Variable pricing	Free to HOVs	Coordinated with bus service	Heavy trucks (more than two axles)	Electronic toll collection	Peak charge per mile (approximate for passenger vehicles)
Time of day	No	Yes	Yes	Yes	$0.25

Maryland's Intercounty Connector (ICC) opened in 2011 after decades of controversy. Originally conceived during the 1950s as a link in the never-built Outer Beltway circling the Washington, D.C., metropolitan area, the toll road provides the first expressway connection for the suburbs in the northern parts of Montgomery and Prince George's counties to points west and north, including to the Baltimore/Washington International Thurgood Marshall Airport. Despite the ICC's being an entirely new road,

Maryland Transportation Authority officials adopted a value-pricing approach to setting toll rates.

Tolls on the six-lane expressway (three lanes each direction) vary by time of day, although the ICC has yet to experience any of the daily gridlock that is common on the other expressways in the region. For passenger cars, during peak hours the toll is $0.25 per mile, which drops to $0.20 off peak and to $0.10 for trips during overnight hours. Peak-hour

For daily shopping and services, the VMT tax scenario creates incentives to minimize driving by conducting shopping trips on the way home from other trips (trip-chaining), consolidating shopping trips (once a week instead of three times a week), patronizing "one stop" or "park once" retail districts, shopping online, conducting good and price comparisons online, and visiting shops and restaurants within walking distance of the home or workplace. With the exception of trips to retail development within walking distance of home or work, the number of shopping trips is expected to decline, although the amount spent per trip is expected to increase. Within the aforementioned 20-year time frame, the VMT tax scenario alone is unlikely to lead to significant industry-wide changes in retail practices or development forms, but specific retail locations will be affected.

COURTESY OF THE ICC PROJECT

Although the ICC does not currently experience congestion, the Maryland Transportation Authority adopted a value-pricing approach that varies tolls by time of day. State government also arranged for new bus routes to travel the toll road.

tolls for heavy trucks range from $0.60 per mile for three axles to $1.50 per mile for six axles.

News features regularly investigate why the ICC seems so empty, with some speculating that it is the expense of the tolls, and with others citing the speed limit and the heavy police enforcement. Without congestion as a guide, setting toll rates is more art than science, or as one commenter reflected: "Really, the prices are a mind game; lower it just five cents a mile and I'd probably use it a lot more than I do now."

"Really, the prices are a mind game; lower it just five cents a mile and I'd probably use it a lot more than I do now."

—Washington, D.C.–area resident in reference to the ICC

However, whether the ICC, as a value-priced toll road, is attracting the predicted number of vehicles per day does not matter as much as whether it is on track to meet its revenue projections. As of October 2012, toll revenues reportedly were running $1 million ahead of projections. While there is little talk of lowering the tolls, raising the 55-mile-per-hour posted speed limit to 65 is under discussion.

The state government arranged for five new bus routes to use the ICC. The airport route, with connections to Washington's Metro rail network, runs hourly seven days a week and is proving a popular bus option for airport travelers. The other four routes provide weekday commuter service with seven to nine trips per day. The ICC does not itself host infrastructure specifically designed for transit use.

Because it was a long-planned highway, land use patterns in Montgomery and Prince George's counties likely adjusted for the ICC for decades before it was built. Except for two development sites near the ICC's eastern terminus, there has been little public discussion of coordinating the opening of the ICC with intensified development along the corridor. The economic development benefits identified in studies are mostly tied to the significant travel time savings to existing hubs of economic activity.

For most retail locations, if the VMT tax scenario affects their attractiveness to customers, it will also affect the retail location's ability to attract and retain workers. Difficulties in attracting workers would start to materialize first in locations that are auto-dependent and isolated from pools of retail workers, such as high-end retail or restaurant districts in wealthy suburbs. In such situations, retail employers may have problems similar to those faced by industrial employers when it comes to finding and retaining workers.

Significance of impacts. Study participants concluded that, of the transportation revenue mechanisms examined, the VMT tax scenario is the one most likely to affect the greatest number of travel and location decisions within metropolitan areas. Even with the relatively high tax rate, the VMT tax scenario's impact is still modest. Other factors such as the cost of fuel, the cost of parking, the availability of transportation alternatives, the desirability of destinations, and travel time are still likely to hold most of the weight when people make trip and location decisions.

For some people, however, the VMT tax scenario will be the factor that tips the scales in favor of the closer destination, the forgone trip, or the location with transportation alternatives. For those already attracted to or willing to consider living, working, or playing in a compact, mixed-use, walkable node, the VMT tax scenario will give them another reason to choose walkable locations. But for individuals who have no desire to spend time in mixed-use, compact, walkable nodes, the VMT tax scenario probably will not raise the cost of driving high enough to change their minds. They will either spend the extra money to continue to travel as before or cut back on trips.

In the absence of some sort of government policy intervention, individuals with lower incomes are the most likely to respond to the VMT tax scenario by forgoing trips, especially far-flung trips, and to find themselves priced out of living in or near compact, mixed-use, walkable nodes.

Variable VMT tax rates. The VMT tax scenario asked study participants to consider a flat tax applied equally to every mile driven. Several study participants observed that some of the downsides to the VMT tax scenario could be mitigated with a variable VMT tax. A variable VMT tax could be adjusted by time of day, location, type of trip, or driver income level. (For example, certain eligible workers could register their work trip routes and get a discounted VMT tax rate for registered trips or a tax credit on their income taxes for total registered commute miles.) Others suggested that the government could charge lower rates for rural travel (where trips tend to be longer) and higher rates for urban travel (where trips tend to be shorter).

Whether and how to vary VMT tax rates will be public policy decisions. If VMT tax rates are varied by geography—such as rural versus urban rates, city versus county rates, suburb versus city rates, etc.—how to vary the VMT tax rates will also have land use implications.

Scenario 5: Toll the Interstate Highway System

In the toll road scenario, study participants were asked to consider what would happen if the Interstate Highway System (IHS) were converted to toll roads. In this scenario, all IHS lane-miles are tolled; there are no free lanes. All future additions to the IHS, including new segments and added lane capacity, are also tolled. The scenario used a flat toll, charged per mile. While tolling the IHS would have potentially significant impacts on intercity freight and passenger travel and interact with land use in rural areas, the scenario asked study participants to focus on impacts on metropolitan travel and land use.

As in the other increased-tax scenarios, the toll road scenario's "tolls" are really taxes on driving. Unlike conventional tolls, the "toll tax" funnels revenue into the general fund and continues to be collected once any debt incurred for the road segment is paid off. Toll revenue from a specific segment of the IHS is not necessarily designated, or even prioritized, for that specific segment of highway.

The tolling scenario was designed to be a proxy for a metropolitan network of high-speed tolled expressways, with the realization that the extent and reach of the IHS differ by metropolitan area. In some metropolitan areas, the IHS is the entire network of high-speed expressways. In other metropolitan areas, state and local expressways supplement the IHS. This variability in the extent and reach of the tolled expressway networks turned out to be an important factor in study participants' discussion of the impacts of the toll road scenario on metropolitan development.

The toll road scenario, like the VMT tax scenario, increases the cost of driving and increases awareness of the cost of driving. Unlike in the VMT tax scenario, where the only way to minimize VMT taxes is to limit driving, under the toll road scenario, taxes are minimized—and the cost of driving reduced—by avoiding toll roads.

Impact on land use: toll road interactions with land use depend on local circumstances. Understanding the potential impacts of tolled expressways on metropolitan land use came down to a series of questions:

- Will tolled expressways attract or repel local, intrametropolitan trips? Which types of trips?
- In the long run, will tolled expressways attract or repel land uses? Which types of land uses?

To answer these questions, study participants observed that the first step is developing a solid understanding of the interaction between the toll road network and the free road network and between the road network and the transit network.

In general, if there are few options to avoid the tolls—i.e., free road options are much slower or nonexistent and there are no good transit alternatives—the toll road scenario will begin to interact with land use like the VMT tax scenario does, only with specific geographic winners and losers. If there are more options to avoid the toll road by taking free roads or transit, the toll road scenario will begin to interact with land use more like the value-priced managed-lane scenarios.

Initially, districts within the metropolitan area that can be easily accessed only via toll roads will be at a competitive disadvantage compared with parts of the metropolitan area that have good access via toll roads, free roads, and transit. This could be particularly disruptive for existing land uses whose primary transportation connection to customers or tenants is converted from a free expressway to a toll road. For example, imagine two office parks dependent on two different expressways, but equally accessible to workers within a metropolitan area. If only one of the expressways is converted to a toll road, the office park on the toll road is now at a competitive disadvantage compared with the office park where workers and deliveries still arrive via the free road. The only mitigating factor for the tolled expressway would be if it now had less congestion.

STATE HIGHWAY 130 EXTENSION, SECTIONS 5 AND 6

AUSTIN–SAN ANTONIO CORRIDOR

Variable pricing	Free to HOVs	Coordinated with bus service	Heavy trucks (more than two axles)	Electronic toll collection	Peak charge per mile (approximate for passenger vehicles)
No	No	No	Yes	Yes	$0.15

"Texas opens a new toll road" is hardly news. Toll roads have been spreading throughout Texas for at least a decade. The August 2012 announcement of the pending opening of a new 41-mile tollway between Austin and San Antonio caught fire in the national media because of its speed limit: 85 miles per hour, the highest posted speed limit in the nation. Providing a longer but higher-speed alternative to the congested I-35, the new State Highway 130 runs through farmland, ranch land, and the small communities of Lockhart (population 12,800) and Seguin (26,000). Local planning documents forecast an additional 70,000 acres of development associated with the toll road.

The toll road is the product of a public/private partnership between the Texas Department of Transportation (TxDOT) and the SH 130 Concession Co., with a majority ownership stake held by Cintra, a Spanish company. Texas retains formal ownership of the road, but the company was responsible for financing, designing, and constructing the facility and has a 50-year contract for operations and maintenance. Risks and revenues are shared between Texas and the company, although the State Highway Fund is on the hook if the facility underperforms. TxDOT's share of the revenues rises as travel grows according to a complicated formula; there is no cap on the company's rate of return. Tolls are currently flat, per-mile charges that do not vary by time of day. Heavy trucks pay about five times the rate of passenger cars, or about $0.75 per mile.

The 85-mile-per-hour maximum travel speed is not a guarantee against future congestion-induced slowdowns. The company, however, is responsible for maintaining minimum speeds, and appears to be on the hook for additional capacity if, over three

consecutive months, more than 10 percent of hourly speeds are less than 60 miles per hour for more than 10 percent of the time or less than 65 miles per hour for more than 5 percent of the time. It is not clear whether the company may use variable tolls to keep speeds above these thresholds.

Such a massive investment in highway capacity in a fast-growing region inevitably leads to Texas-sized booster dreams. For Lockhart, located midway between Austin and San Antonio, State Highway 130 reduces travel time from Lockhart into Austin and San Antonio by about 15 to 20 minutes. Lockhart and Caldwell County leaders are hoping that the toll road is the magic ingredient that will finally elevate this rural community known

For Lockhart, a rural community halfway between San Antonio and Austin, hopes rise that the new toll highway will bring new development into this fast-growing region.

COURTESY OF THE SH 130 CONCESSION COMPANY LLC

for its barbecue to the status of official Austin suburb and bedroom community. As quoted in the Austin *Statesman*, Caldwell County Commissioner John Cyrier outlined Lockhart's plans: "We're trying to attract a business. Then the rooftops will come, and the retail. We're hoping for at least 10,000 rooftops." In preparation for the toll road, Lockhart updated its comprehensive plan, which calls for annexing land and enhancing regulations to managed development along SH 130.

> *"We're trying to attract a business. Then the rooftops will come, and the retail. We're hoping for at least 10,000 rooftops."*
>
> —Caldwell County Commissioner John Cyrier

At an August 2012 local development conference, Javier Gutierrez, the CEO of the SH 130 concession company, reflected the local enthusiasm for expanded development: "We really believe this area is going to explode in the next year." Websites are already advertising thousands of acres for residential, industrial, and mixed-use development.

Over the very long run, locations dependent on the toll road may find their own competitive niche. As expressways, the toll roads still offer speed. Although land users with a lower value on speed are likely to avoid sites dependent on the toll road, land users with a higher value on and who can pay for speed may find toll road locations attractive.

If the toll road is just an additional travel option, and land can still be accessed via free roads and transit, then the toll road can be thought of as an amenity that some drivers find worth paying for some of the time. If one assumes that the toll road provides a better travel experience—higher speeds and reliability—land uses associated with businesses and residents with a high value on, and who are able to pay for, these amenities will be attracted to sites served by toll roads.

Discussion among study participants only touched on which types of land uses would be attracted or repelled by toll roads. How land use reacts would be different in a metropolitan area where all the roads providing expressway service are tolled versus in a metropolitan area where there is a mix of tolled and free expressways. In the former scenario, locating near the tolled expressway is more likely to be an economic decision based on business needs or family budgets. In the latter, the choice of locating near a tolled expressway will be similar to buying a luxury good, and the amenities gained through paying tolls would compete with other uses for discretionary spending.

Conclusion

The ULI Infrastructure Initiative constructed and conducted a thought experiment on the interaction among tolling, value pricing, and land use as a way to raise awareness of and frame future research on the intended and unintended consequences of the spread of toll roads and value-priced, managed lanes in the United States. The thought experiment brought together leaders from the transportation and land use fields to discuss the impacts on land use in metropolitan areas over a 20-year time frame. The study also looked at how the interaction of toll roads, value-priced lanes, and land use would potentially affect lower-income groups.

The thought experiment yielded fruitful discussions about the significance of impacts; the difference between tolling, VMT taxes, and value pricing; and the relationship between transportation revenue mechanisms and trends favoring development in compact, mixed-use, walkable nodes. Study participants concluded that the impacts will be real, but modest, and will vary greatly by metropolitan area. The VMT tax is likely to have the most widespread impacts. The land use impacts of tolling and value-priced lanes are more likely to appear on the corridor level. For lower-income groups, participants in the study identified the most significant areas of concerns as the connection between workers and jobs and the affordability of locations served by managed lanes connecting to compact, mixed-use, walkable nodes.

PART 3:
Conclusions for Today's Decision Makers

Surveying the changes that are coming to metropolitan transportation networks can be exciting, but daunting. Metropolitan areas are already transforming their transportation networks. New value-priced, congestion-free highway lanes continue to open. Toll roads bring new road capacity or put a price on travel to pay for reconstructing old roads and bridges. Communities debate whether to improve intraregional and commuter bus service. All these changes join the expanded and planned investment in light-rail, commuter-rail, and local transit services, and the growing popularity of locating in compact, mixed-use, walkable nodes.

Travelers will have more options, but policy makers, implementers, and developers also will have more choices. The careful consideration required before making both transportation and land use investment decisions will become more complicated. The scenarios examined in this study help explore how decisions about collecting revenue from transportation interact with land use, but they do not take into account political realities and implementation feasibility. In addition, the revenue mechanisms vary in how quickly they are likely to be implemented and to begin to interact with land use in the United States.

Increasing taxes on motor fuels, while not easy from a political perspective, is a tried-and-true revenue mechanism. As compared with tolling and taxing VMT, taxing motor fuel has the least potential to alter land use decision making, and it also provide no new amenities for land development or mobility. VMT taxes or fees—flat and variable—raise significant issues for land use. Unproven technologies and political resistance in the United States, however, mean that VMT taxes or fees are not likely to achieve widespread implementation anytime soon.

Leaders in Maryland celebrate the opening of the Intercounty Connector.

The use of tolling and value pricing, on the other hand, is expanding in the United States. These two strategies are already interacting with land use decisions, and policy decisions being made today about how they are built and operate will soon play out on the American landscape.

A Broader Look at Toll Roads

The study's toll road and managed-lane scenarios hold the risk of paying attention to the raindrops (tolls), while ignoring the roaring river (new highway lanes). In the real world, Americans' apparent willingness to pay tolls is seen as a way to break through the capital investment roadblock caused by Americans' unwillingness to raise taxes on motor fuels.

New highway capacity has the potential to significantly influence land use. Toll roads are being planned and built that will add significant capacity to the highway network, either through converting arterials and rural highways to expressways or through building entirely new roads on new alignments. The new highway capacity and new alignments have just as much, if not more, potential to interact with land use than just the fact that these new roads or added lanes are tolled.

In addition, flat tolls will start to be seen as limiting and obsolete. The managers of both new and existing toll lanes/roads will explore and then embrace value pricing to provide user amenities such as free-flowing travel and to maximize revenue. As discussed in the managed-lane section, value pricing involves trade-offs that are ultimately policy decisions that will be decided in the public policy arena. Because the amenities associated with value pricing require that tolls be a permanent feature unrelated to the cost of the road (and not a temporary measure to pay off the construction bonds), public policy discussions also need to focus on appropriate uses for the excess revenue.

Expanding the use of tolling is also an avenue for increasing private sector investment in infrastructure, further complicating the already complex interaction between tolled lanes/roads and land use. Bringing private investors into the process adds additional stakeholders—and interests—to policy decisions over how the roads will operate and the use of excess revenue. Private investors in infrastructure may also seek opportunities to coordinate their transportation investments with investments in land development, taking advantage of the transportation amenities to capture increases in land values or push for their own development proposals.

For land developers and investors, the spread of value-priced toll roads brings new risks as well as new opportunities. Tenants may be fickle, being attracted to locations near toll road amenities in good times and demanding discounts in bad times, and developers and investors will need to become more aware of how different market segments are likely to respond to tolls. The tolls themselves may be unpredictable and fast-changing, creating even more uncertainty. For sites dependent on access via toll roads, property managers and employers may discover that they need to pay much more attention to transportation costs and to attract tenants, buyers, and workers with transportation alternatives or subsidies or provide other compensating amenities.

Transportation Revenue Mechanisms and Equity

This study has added the land use dimension to discussions about transportation revenue mechanisms and social equity, finding that the most significant areas of concerns are the connection between workers and jobs and the affordability of locations in compact, mixed-use nodes or served by managed lanes. It should be recognized, however, that the prism of land use is only one way to look at how changing how Americans pay for transportation affects societal equity. Policy decisions such as whether to direct the revenue from other taxes—such as sales or income taxes—to transportation may be equally significant, or more significant, to the welfare of lower-income groups.

Simply increasing the cost of driving without providing new benefits will obviously hit lower-income groups dependent on driving harder than other income level groups. If the country increases taxes on motor fuels, lower-income groups have less ability to buy new fuel-efficient vehicles. If the country shifts to VMT taxes, lower-income groups will find themselves more quickly priced out of the close-in nodes that minimize driving. If the country shifts to tolled expressways, lower-income groups will be the first ones to opt for the slow, free roads or to choose to avoid destinations because of the expense of the tolls. In the long run, a metropolitan area that sorts its land uses by people's value of and ability to pay for high-speed travel is a metropolitan area that becomes more segregated. In the absence of government policies to balance these impacts, lower-income groups will lose out.

Fortunately, as this report has shown, it is possible to structure these transportation revenue mechanisms to provide new benefits. Instituting value pricing on managed lanes and on toll roads has the potential to provide improved mobility and accessibility to a broad segment of the population, but only if implemented with this objective in mind. Public policy decisions about how the value-priced roads and associated transit services are planned, designed, and operated and how any excess revenues are spent will decide whether society benefits broadly or narrowly.

Implications for Developers, Planners, and Researchers

With the nation's limited experience implementing these revenue mechanisms and coordinating them with land use decisions, the professional communities of developers, planners, and researchers have much to learn, but also much to contribute to the wider policy and decision-making conversation.

Developers, planners, and researchers should work within their respective professional communities to advance and disseminate knowledge about the new uses for value pricing, tolling, and taxing VMT and their impact on land use. In addition, reaching out across disciplines—cross-fertilization among developers, planners, and researchers— will strengthen efforts to conceptualize research projects, develop best practices, and set standards. Just as important, developers, planners, and researchers offer unique perspectives that will be valuable to policy makers as they forge federal, state, and local policies; pursue visions, implement programs, and design projects; and strive to ensure that new transportation services and the land uses that are attracted to them benefit a broad swath of the metropolitan population.

In addition, realizing the potential of these revenue mechanisms to support—instead of hamper—growing market preferences for development in compact, mixed-use, walkable nodes will take proactive and concerted action. Innovations in land use planning, urban design, and the development process are needed, and thoughtful discussions about how to use the revenue generated by these facilities will be required. Key areas for innovation include the integration with high-quality transit service and the coordination with development in compact, mixed-use, walkable nodes.

Bus-oriented, mixed-use development in Broomfield, Colorado, includes a 6,000-seat entertainment center, townhouses, apartment communities, and a new main street with retail and office space.

Conclusion

As the United States expands the use of tolling and broadens the reach of value-priced, congestion-free lanes, this ULI Infrastructure Initiative effort was designed to raise awareness of land use. Local policy makers and the land development community need to develop a more robust understanding of how tolling and value pricing may influence land use in their communities. They should work with state and federal policy makers to help integrate land use considerations into the planning, design, and implementation of these new roadways and congestion-free lanes. Although these transportation revenue mechanisms have the potential to bring new benefits to residents and businesses in metropolitan areas and to support development in compact, mixed-use, walkable nodes, fully realizing these benefits will require the concerted effort of policy makers, researchers, planners, and the development community.

Acknowledgments

The ULI Infrastructure Initiative gratefully acknowledges the Rockefeller Foundation for its support of this report and related activities.

The views, opinions, and recommendations expressed in this report are the responsibility of the ULI Infrastructure Initiative and do not necessarily reflect those of individual ULI members, individual participants in the workshops and interviews conducted as part of the research project, or the project funder.

List of Participants

The ULI Infrastructure Initiative appreciates the contributions of the following people who participated in the workshops and interviews conducted as part of this research project:

Asha Weinstein Agrawal
San Jose State University

Nat Bottigheimer
Washington Metropolitan Area Transit Authority

Robert Brodesky
IHS Consulting

Kenneth R. Buckeye
Minnesota Department of Transportation

Robert Burke
Greenheart Land Company

Susanne E. Cannon
DePaul University Real Estate Center

Denise M. Casalino
AECOM

Steve Crosby
CSX

John Q. Doan
Atkins North America

Robert Dunphy
Transportation Consultant

Tyler Duvall
McKinsey

William Kohn Fleissig
Communitas Development

Emil H. Frankel
Bipartisan Policy Center

Stephen B. Friedman
SB Friedman and Company

Ron Golem
Bay Area Economics

Ginger D. Goodin
Texas Transportation Institute

Doug Johnson
MTC–Metropolitan Transportation Commission

Kevin J. Krizek
University of Colorado

Todd Litman
Victoria Transport Policy Institute

Deron Lovaas
Natural Resources Defense Council

Gerard C.S. Mildner
Portland State University

Robert Palmer
RS&H

Alexander Quinn
AECOM

Sandra Robles
ULI Terwilliger Center for Housing

Bob Romig
Florida Department of Transportation

Alex Rose
Continental Development Corporation

Sandra Rosenbloom
Urban Institute

Lynn Ross
ULI Terwilliger Center for Housing

Joshua L. Schank
Eno Center for Transportation

Elizabeth M. Seifel
Seifel Consulting Inc.

Samuel N. Seskin
CH2M HILL

Elliot Stein
ULI San Francisco

Yaromir Steiner
Steiner + Associates Inc.

Brian D. Taylor
University of California at Los Angeles

Susanne Trimbath
STP Advisory Services LLC

Jeff Weidner
Florida Department of Transportation

Kate White
San Francisco Foundation

Shirley Ybarra
Reason Foundation

Special Thanks

The ULI Infrastructure Initiative extends special thanks to the following:

Robert Palmer and Alexander Quinn for chairing the workshops in Washington, D.C., and San Francisco, respectively.

Marilee Utter, Emil Frankel, Joshua Schank, Martine Micozzi, and Asha Weinstein Agrawal for their review of and assistance with constructing the thought experiment and advice on participants.

Gideon Berger and Loren Bloomberg for their advice and guidance on technical aspects of this research project.

Ed McMahon, Trey Davis, Caroline Sullivan, and Anita Kramer for their assistance with the workshop in Washington, D.C., and Kate White, Wendy Tao, Torrey Wolff, Jason Bernstein, and Sheppard Mullin for their assistance with the workshop in San Francisco.

John Doan and David Ungemah for the invitation to present an early version of this work to the Transportation Research Board Pricing Committee in June 2012.